International Conference on Biomolecular Engineering (ICBE Asia 2018)

Singapore
8-10 January 2018

ISBN: 978-1-5108-9198-2

International Conference on Biomolecular Engineering (ICBE Asia 2018)

Singapore
8-10 January 2018

AIChE
The Global Home of Chemical Engineers

Printed from e-media with permission by:

Curran Associates, Inc.
57 Morehouse Lane
Red Hook, NY 12571

CURRAN ASSOCIATES INC.
proceedings.com

Some format issues inherent in the e-media version may also appear in this print version.

Printed by Curran Associates, Inc. (2019)

AIChE
120 Wall Street, FL 23
New York, NY 10005-4020

Phone: (800) 242-4363
Fax: (203) 775-5177

www.aiche.org

Additional copies of this publication are available from:

Curran Associates, Inc.
57 Morehouse Lane
Red Hook, NY 12571 USA
Phone: 845-758-0400
Fax: 845-758-2633
Email: curran@proceedings.com
Web: www.proceedings.com

Sessions:

8-Jan-18

Papers:

Session 4: Biomolecular Programming - from DNA to Community

Poster Session

Plenary Talk: Glycosylated cationic block poly(?-peptides) for potentiating antibiotics against Gram-negative bacteria

Conference: International Conference Biomolecular Engineering ICBE

Year: 2018

Proceeding: ICBE Asia 2018: International Conference on Biomolecular Engineering

Group: General Submissions

Session: Plenary Lecture: Mary Chan, Nanyang Technological University

Time: Monday, January 8, 2018 - 9:00am-10:00am

Authors:

Chan, M., Nanyang Technological Univerisity

Zhangyong Si[1], Yu Du[1], Moon[1], Damien[1], Jothy[1], Yang Liu[2], Kaixi Zhang[1], Sheethal[1], Chun Kiat Yeo[1], Yuguang Mu[2], Kevin Pethe[3], MB Chan-Park[1]*

1. Centre for Antimicrobial Bioengineering, School of Chemical and Biomedical Engineering, NTU, 62 Nanyang Drive, Singapore 637459, Singapore.
2. School of Biological Sciences, NTU
3. Lee Kong Chian School of Medicine, NTU

The rise of multi-drug resistant (MDR) bacteria together with the decrease of useful antibiotics has been a mounting problem. Especially for Gram-negative bacteria, no new family of drug was found in the last past 50 years. Herein, we developed a novel glyco-Î²-peptides through one-step polymerization, following by one-pot global deprotection to get water soluble polymers. Our glyco-Î²-peptides had relatively good antimicrobial activity against both Gram-negative *P. aeruginosa* and Gram-postive *S.aureaus*, with MIC of 64 µg/mL. In vivo biocompatibility test using a murine intravenous toxicity model at dosage of 10 mg/kg indicated that glycopeptide had excellent biocompatibility, with 100% animal survival after 7 days and no significant difference in all the clinical important biomarker compared to the control groups. Our polymer could work synergistically with antibiotics at low concentration (16µg/mL), to potentiate conventional traditional antibiotic against wide type*P.aeruginosa* and their clinical relevant strain PAN-sensitive (PAES), clinical isolate P. aeruginosa D25, P. aeruginosa D1 and clinical multi-drug resistant strain PAER (almost no antibiotic showed effect except colistn) greater than 16 times.

Invited Talk: Programmable synthetic gene circuit as a potential therapeutic intervention for liver cancer

Conference: International Conference Biomolecular Engineering ICBE

Year: 2018

Proceeding: ICBE Asia 2018: International Conference on Biomolecular Engineering

Group: General Submissions

Session: Session 1: Foundational Technologies for Biomolecular Engineering

Time: Monday, January 8, 2018 - 10:30am-11:05am

Authors:

Xie, Z., Tsinghua University

Oncolytic viruses are genetically engineered or naturally occurring viruses that selectively replicate in cancer cells without damaging the normal cells. The replication of oncolytic virus can lysis cancer cells and trigger immune responses against cancer cells. However, it remains a great challenge to improve the specificity and efficacy of oncolytic virus in a predictable and programmable manner. Synthetic gene circuits are designed to implement desired cellular functions in living cells through sensing, integration and processing of molecular information. Here, we developed two types of sensory switches by using synthetic TALE repressors and deactivated Cas9 (dCas9) repressors, which can response to cancer specific microRNAs. We demonstrated that using the input-output transfer curves of genetic parts enables accurate prediction of the behavior of modularly assembled switch circuits. We also showed that TALER and dCas9 switches employing feedback regulation exhibit improved accuracy for microRNA-based cancer cell classification. In addition, we cloned the sensory switch circuit into an adenoviral vector to control the virus replication. We demonstrated that the engineered adenovirus can be used to specifically kill hepacellular carcinoma cells in vitro and in the mouse model. Our results provide a novel strategy to program and improve the specificity of oncolytic virus replication in target cancer cells, which may lead to innovative therapeutics for cancers.

Formatting Issue - Best Available Paper/Graphic

Lost in Translation: Mapping Ribosomal Active Site Mutations in Vitro

Conference: International Conference Biomolecular Engineering ICBE

Year: 2018

Proceeding: ICBE Asia 2018: International Conference on Biomolecular Engineering

Group: General Submissions

Session: Session 1: Foundational Technologies for Biomolecular Engineering

Time: Monday, January 8, 2018 - 11:05am-11:30am

Authors:

d'Aquino, A., Northwestern University
Azim, T., Northwestern University
Hockenberry, A. J., Northwestern University
Aleksashin, N., University of Illinois at Chicago
Mankin, A., University of Illinois at Chicago
Jewett, M. C., Northwestern University

The ribosome, a 2.5-MDa molecular machine that polymerizes Î±-amino acids into proteins, is the catalytic workhorse of the translation apparatus. The catalytic capacity of the translation machinery has attracted extensive efforts to repurpose it for novel functions. One key idea is that the natural translation machinery can be harnessed to synthesize polymers containing non-natural building blocks. Expanding the repertoire of ribosomal substrates and functions is a difficult task, however, because the requirement of cell viability severely constrains the alterations that can be made to the ribosome, a catalyst that sustains the life of a cell. In practice, these constraints have made the natural ribosome nearly unevolvable and, so far, no generalizable approach for modifying the catalytic peptidyl transferase center (PTC) of the ribosome has been advanced. We propose to address this grand challenge by using cell-free systems that harness the biosynthetic potential of cellular machines without using intact cells, thus removing cell viability constraints.

Here, we use our in vitro ribosome synthesis, assembly, and translation system (termed iSAT) to generate variant ribosomes with mutations in the active site, and inquire how these modifications impact protein synthesis. The active site of the ribosome is composed solely of ribosomal RNA (rRNA). This rRNA is assembled into three key loops, which facilitate translation: the peptidyl transferase loop, the A-loop, and the P-loop. The peptidyl transferase loop plays a key role in positioning substrates, is involved with peptide stalling, and is a key target for antibiotic binding. The A- and P-loops flank the peptidyl transferase loop within the ribosomeâ€™s active site. The A-loop is responsible for aminoacyl-tRNA interactions with the ribosome, and the P-loop is known to play a role in further facilitating peptide bond formation between amino acids. Because mutations to many of the nucleotides in these three active site loops confer lethal phenotypes, few to no studies exist in probing and fully characterizing them.

Understanding the effects of single-base mutations on these active site nucleotides will provide insight into the basic biochemistry of these bases in translation, but also provide the groundwork for engineering the catalytic center of the ribosome. Using iSAT, we have assembled 180 different variant ribosomes possessing single-base substitutions of 23S rRNA nucleotides in these active site loops. By successfully quantifying full-length protein synthesis kinetics of iSAT-assembled wild type and mutant ribosomes, we unexpectedly found many key PTC mutations, which were expected to abolish ribosomal activity, still permitted full-length protein synthesis. We also assessed translation fidelity and ribosome assembly, as well as mapped mutant activity onto the ribosomeâ€™s crystal structure. Our work provides the first and only comprehensive mapping of the impacts of every mutation within the ribosomeâ€™s active site on protein synthesis. The understanding gained from these studies facilitates efforts to engineer and evolve ribosomes for synthetic biology.

Synthetic Decoupling of Transcription and Translation Processes for the Quality Control of Gene Expression in Escherichia coli

Conference: International Conference Biomolecular Engineering ICBE

Year: 2018

Proceeding: ICBE Asia 2018: International Conference on Biomolecular Engineering

Group: General Submissions

Session: Session 1: Foundational Technologies for Biomolecular Engineering

Time: Monday, January 8, 2018 - 11:30am-11:55am

Authors:

Seo, S. W., Seoul National University

Tightly coupled transcription and translation system in prokaryotes can cause the production of non-functional polypeptides which are translated from early-terminated or degraded transcripts. In this study, a synthetic gene expression cassette was designed to improve the yield of the synthesis of full-length proteins in *E. coli*. We used toehold switches to decouple transcription and translation processes. By placing corresponding trigger sequences at 3â€™-end of mRNA, intact full-length mRNAs could only be translated. When fluorescence tags were attached at both N- and C-terminal ends, we found that equivalent fluorescent intensities from both ends were observed in our system. This result shows that decoupling of transcription and translation processes can be applicable to elevate the quality of gene expression, especially longer proteins.

Formatting Issue - Best Available Paper/Graphic

Modeling Gene Circuit Expression Dynamics in Cell and Cell-Free Systems

ICBE Asia

Conference: International Conference Biomolecular Engineering ICBE

Year: 2018

Proceeding: ICBE Asia 2018: International Conference on Biomolecular Engineering

Group: General Submissions

Session: Session 1: Foundational Technologies for Biomolecular Engineering

Time: Monday, January 8, 2018 - 11:55am-12:20pm

Authors:

Yeoh, J. W., Life Sciences Institute
Jayaraman, P., National University of Singapore
Poh, C. L., National University of Singapore

A detailed understanding of underlying gene circuit expression dynamics forms the basis towards achieving optimized system performances. Despite a growing wealth of experimental tools, it remains challenging to measure the reaction kinetics underlie the system of interest. Computational modelling provides an invaluable alternative to complement the experiment in a more time and cost-effective quantitative manner. These mathematical approaches, spanning from phenomenological to biophysical, discrete to continuous, stochastic to deterministic, decipher the principles govern the underlying functions and dynamics of various system components. As a case in point, in recent years, the advances in optogenetic systems enable one to control cells with unprecedented spatiotemporal precision that could be intractable using conventional chemical-triggered regulators. We have recently developed a blue light and chemical-controllable promoter system which allows rapid and reversible control of chemically inducible promoters, resolving the difficulty of achieving reversible control suffered by traditional chemical inducible systems. To fully harness the mechanistic properties of these hybrid systems, we employed an *in silico* kinetic modeling approach in an attempt to bridge the gap between the genotype and the observed dynamics of the two dual-input promoter systems. Our biophysically-based kinetic model accounts for the essential reactions involved in the functioning system. Indeed, the model was able to recapitulate experimental data and capture some key features, including concentration-dependent delay responses for inducible expression and blue light repression, and delay observed in temporal control behaviors. It was also inferred from the model that the affinity of blue light repression is inducer concentration-dependent, in which high inducer concentration could reduce the repression affinity. To improve the system performance, further model analyses were conducted. In parallel, we also applied the deterministic mathematical modelling approach in conjunction with experimental characterization to study an engineered one component-based blue light inducible promoter in the cell-free environment. As a proof of concept, our model was able to describe the system behaviors quantitatively and predict a blue light ON-OFF illumination pattern which allows us to generate an oscillatory waveform. These tandem experiment/modelling approaches expedite the iterative â€˜design-build-testâ€™ cycles, advancing our ability towards rapid prototyping of complex biological systems.

How do Computational Approaches Benefit Life Science Research? Case studies into immune, cancer and developmental cell dynamics

ICBE Asia

Conference: International Conference Biomolecular Engineering ICBE

Year: 2018

Proceeding: ICBE Asia 2018: International Conference on Biomolecular Engineering

Group: General Submissions

Session: Session 1: Foundational Technologies for Biomolecular Engineering

Time: Monday, January 8, 2018 - 12:20pm-12:45pm

Authors:

Selvarajoo, K., BioTrans, A*STAR Singapore

Systems biology efforts are revealing complexities far beyond traditional reductionist approaches can anticipate. Single cells display stochastic or noisy gene/protein expressions and, sometimes, show emergent response when pooled into populations. Should modern day research move towards single cell research, as proposed by â€œsingle-cell biologistsâ€, or continue to explore the dynamics of tissue-level response? Our research, over a decade, has focused on dynamic cellular behaviors to uncover novel network and global response properties both at single cells and populations. We investigated instructive cell signaling and high-throughput transcriptome-wide behaviors of immune, cancer, and embryonic development cells. Our data reveal that cultured populations display deterministic signaling that can be modeled using *linear response rules*. Adopting the *rules*, we have computationally predicted and verified experimentally novel signaling features and key targets for controlling proinflammatory responses and cancer survival. In addition, our transcriptome-wide statistical analyses of single cell embryonic developmental process, from oocyte to blastocysts cell stage differentiation, revealed global mechanisms for the onset of noisy transcription responsible for generating gene expression variability.

Selected References

1. Selvarajoo K. A systems biology approach to overcome TRAIL resistance in cancer treatment. *Prog Biophys Mol Biol.* 2017, 128:142-154.
2. Piras V, Selvarajoo K. The reduction of gene expression variability from single cells to populations follows simple statistical laws.*Genomics*. 2015, 105:137-44.
3. Piras V, Tomita M, Selvarajoo K. Transcriptome-wide variability in single embryonic development cells. *Sci Rep.* 2014, 4:7137.
4. Hayashi K, Piras V, Tabata S, Tomita M, Selvarajoo K. A systems biology approach to suppress TNF-induced proinflammatory gene expressions. *Cell Commun Signal.* 2013, 11:84.
5. Tsuchiya M, Piras V, Choi S, Akira S, Tomita M, Giuliani A, Selvarajoo K. Emergent genome-wide control in wildtype and genetically mutated lipopolysaccarides-stimulated macrophages.*PLoS One*. 2009, 4:e4905.

Invited Talk: Lactic acid production using C1 or C6 carbon source

Conference: International Conference Biomolecular Engineering ICBE

Year: 2018

Proceeding: ICBE Asia 2018: International Conference on Biomolecular Engineering

Group: General Submissions

Session: Session 2: Translational Biomolecular Engineering

Time: Monday, January 8, 2018 - 2:00pm-2:35pm

Authors:

Hahn, J. S., Seoul National University

Biorefinery is defined as producing useful chemicals and fuels mainly via microbial fermentation. Currently, glucose and other biomass-derived sugars are used as major carbon sources for microbial fermentation, but other abundant and inexpensive carbon sources such as CO_2, CO and methane are considered as promising alternatives to sugars in biorefinery. We engineer *Saccharomyces cerevisiae* and *Methylomonas*sp. DH-1 to produced D-lactic acid using glucose and methane, respectively. Both organisms cannot naturally produce lactic acid. Therefore L or D form-specific lactic acid can be produced by introducing stereo-specific heterologous lactate dehydrogenase. By eliminating ethanol and glycerol production pathways, and adaptive laboratory evolution to increase lactate tolerance, we generate *S. cerevisiae* strain, which can produced D-lactic acid under acidic conditions with high titer, yield, and productivity. *Methylomonas* sp. DH-1 is a methanotroph which can use methane as a sole carbon and energy source. By adaptive laboratory evolution, lactic acid-tolerant strain is generated. Expression of D form-specific lactate dehydrogenase gene in the evolved strain leads to about 8-fold increase in D-lactic acid production from methane compared to that in wild type strain.

Formatting Issue - Best Available Paper/Graphic

Effect of Alkylation on the Cellular Uptake of Polyethylene Glycol-Coated Gold Nanoparticles

ICBE Asia

Conference: International Conference Biomolecular Engineering ICBE

Year: 2018

Proceeding: ICBE Asia 2018: International Conference on Biomolecular Engineering

Group: General Submissions

Session: Session 2: Translational Biomolecular Engineering

Time: Monday, January 8, 2018 - 2:35pm-3:00pm

Authors:

Choi, C. H. J., California Institute of Technology

Alkyl groups (CnH2n+1) are prevalent in engineered bionanomaterials used for many intracellular applications, yet how alkyl groups dictate the interactions between nanoparticles and mammalian cells remains incomprehensively investigated. In this work, we report the effect of alkylation on the cellular uptake of densely polyethylene glycol-coated nanoparticles, which are characterized by their limited entry into mammalian cells. Specifically, we prepare densely PEGylated gold nanoparticles that bear alkyl chains of varying carbon chain lengths (n) and loading densities (termed â€œalkyl-PEG-AuNPsâ€), followed by investigating their uptake by Kera-308 keratinocytes. Strikingly, provided a modest alkyl mass percentage of 0.2% (2 orders of magnitude lower than that of conventional lipid-based NPs) in their PEG shells, dodecyl-PEG-AuNPs (n = 12) and octadecyl-PEG-AuNPs (n = 18) can enter Kera-308 cells 30-fold more than methoxy-PEG-AuNPs (no alkyl groups) and hexyl-PEG-AuNPs (n = 6) after 24 h of incubation. Such strong dependence on n is valid for all serum concentrations considered (even under serum-free conditions), although enhanced serum levels can trigger the agglomeration of alkyl-PEG-AuNPs (without permanent aggregation of the AuNP cores) and can attenuate their cellular uptake. Additionally, alkyl-PEG-AuNPs can rapidly enter Kera-308 cells via the filipodia-mediated pathway, engaging the tips of membrane protrusions and accumulating within interdigital folds. Most alkyl-PEG-AuNPs adopt the â€œendo-lysosomalâ€ route of trafficking, but â^¼15% of them accumulate in the cytosol. Regardless of intracellular location, alkyl-PEG-AuNPs predominantly appear as individual entities after 24 h of incubation. Our work offers insights into the incorporation of alkyl groups for designing bionanomaterials for cellular uptake and cytosolic accumulation with intracellular stability.

In Vivo Anti-Bacterial Phage Therapy through Immunological Cloaking

Conference: International Conference Biomolecular Engineering ICBE

Year: 2018

Proceeding: ICBE Asia 2018: International Conference on Biomolecular Engineering

Group: General Submissions

Session: Session 2: Translational Biomolecular Engineering

Time: Monday, January 8, 2018 - 3:00pm-3:25pm

Authors:

Nam, Y. S.
Yu, J. H.

Phage therapy can serve as an alternative treatment to overcome antibiotic resistance, but the use of phage for anti-bacterial treatments is limited by rapid clearance from systemic circulation. Here we introduce immunologically cloaked T7 phage to reduce phagocytosis through the expression of self-peptide, derived from human CD47, on the major capsid. Self-peptide expressing T7 phage (Self-T7) suppresses up to 70% phagocytosis *in vitro* compared to wild-type T7 phage (WT-T7). Real-time *in vivo* image analysis demonstrates that Self-T7 exhibits a markedly longer blood circulation compared to WT-T7. In a mouse model of bacterial infection of intestines, the survival rate of mice is greatly increased by the intraperitoneal or intravenous injection of Self-T7, while WT-T7 only exhibited limited effects on mouse survival. This work highlights the enhanced*in vivo* efficacy of anti-bacterial phage therapy through the immunological cloaking of phage for suppressed phagocytosis and prolonged blood circulation.

Application of Cationic Polymers for In Vivo Antibacterial Applications

ICBE Asia

Conference: International Conference Biomolecular Engineering ICBE

Year: 2018

Proceeding: ICBE Asia 2018: International Conference on Biomolecular Engineering

Group: General Submissions

Session: Session 2: Translational Biomolecular Engineering

Time: Monday, January 8, 2018 - 3:25pm-3:50pm

Authors:

Zheng, H., Nanyang Technological University Singapore
Vikhe, Y., Nanyang Technological University Singapore
Chan-Park, M. B., Nanyang Technological University

In the recent year, the appearance of multiple resistances of the antimicrobial agent toward the different types of pathogenic bacteria is the most prominent threat to the global public health. The invention of antibiotic to the medical field has great impact on human mortality and morbidity. However, the multi drug resistance (MDR) effect of the pathogenic bacterial towards commonly used antibiotics has impeded the treatment of bacterial infections.

In the recent few decades, the Antimicrobial polymers (AMPos) have been widely used as a promising antimicrobial alternative to combat the infectious disease caused by MDR bacterial. Hence, unlike traditional antibiotics, AMPos interact with the bacterial membrane through non-specific electrostatic interaction. This electrostatic mode of interaction offers the AMPos to progressively interact with the bacterial with a less likely hood of bacterial resistance. We have developed numerous cationic polymer solutions that are bactericidal in vivo with a murine wound model.

We have also developed a series of chitosan-based low molecular weight chitosan derivatives. It can kill MRSA in excisional wound model. The mechanism of kill and the solution properties of these low molecular weight chitosan derivatives shall be presented. antibacterial hydrogel based on AMPos that is intrinsically antibacterial and non-leachable.

Invited Talk: New Wine from Old Barrels: Repurposing Biology through Synthetic Biology

ICBE Asia

Conference: International Conference Biomolecular Engineering ICBE

Year: 2018

Proceeding: ICBE Asia 2018: International Conference on Biomolecular Engineering

Group: General Submissions

Session: Session 3: Translational Biomolecular Engineering

Time: Tuesday, January 9, 2018 - 8:30am-9:05am

Authors:

Yew, W. S., National University of Singapore

Synthetic biology is the engineering of biology for purpose. One of the most common techniques in synthetic biology is the repurposing of tools and processes in Nature for a purposeful function. This normally involves the re-wiring of metabolic processes, channelling energy across systems for the purposeful production of therapeutics, nutraceuticals and ingredients, and more recently, in environmental sustainability. In the first part of the presentation, the biosynthesis of polyketides will be discussed. Polyketides are a large class of biomolecules that are naturally produced by bacteria, fungi and plants, and include many clinically important biomolecules with anti-cancer, anti-microbial, anti-oxidant and anti-inflammatory activities. They are biosynthesized from acyl-CoA precursors by polyketide synthases (PKSs), and due to their chemical complexity, are not easily synthesized and structurally manipulated by chemical means. In the second part of the presentation, efforts made in achieving environmental sustainability, allowing our society to progress without effecting environmental pollution and resource depletion, will be illustrated. Electronic waste represents a significant and growing portion of waste in our cities, and thus there is a pressing need to develop sustainable technologies to recycle electronic waste in order to protect our environment and preserve natural resources. Current recycling methods have proved to be environmentally and economically unsustainable; these treatment technologies for electronic waste use strong acids or cyanide that are pollutive in nature. Synthetic biology provides sustainable solutions to address challenges that confront society. Using Nature as inspiration, our synthetic biology-empowered approach will contribute towards Singaporeâ€™s efforts to transition to a green economy, furthering her efforts to become an environmentally sustainable city

Site-Specific Albumination of Therapeutic Proteins for the Prolonged Serum Half-Life In Vivo

Conference: International Conference Biomolecular Engineering ICBE

Year: 2018

Proceeding: ICBE Asia 2018: International Conference on Biomolecular Engineering

Group: General Submissions

Session: Session 3: Translational Biomolecular Engineering

Time: Tuesday, January 9, 2018 - 9:05am-9:30am

Authors:

Kwon, I., Gwangju Institute of Science and Technology

Gout is a common inflammatory arthritis. The number of people affected by gout reached three millions in US in 1999. Gout is caused by abnormally high concentration of insoluble uric acid in the blood. One way to treat gout is conversion of uric acid into a soluble metabolite by urate oxidase enzyme. PEGylated urate oxidase was approved by FDA for gout treatment. However, immune concerns associated PEG molecules drove search for alternative way to prolong the serum half-life. We used site-specific incorporation of a clickable non-natural amino acid and click chemistry to site-specifically conjugate a human serum albumin to urate oxidase (Uox-HSA). Uox-HSA exhibited the significantly longer serum half-life in animal studies without compromising the therapeutic activities upon HSA conjugation. Using similar techniques, fatty acid, a natural HSA ligand, was site-specifically conjugated to Uox for the enhanced HSA binding affinity without a substantial loss in the therapeutic activities. Furthermore, injectable hydrogels conjugated to HSA allowed the sustained release of fatty acid-conjugated Uox in vivo.

Formatting Issue - Best Available Paper/Graphic

Synthetic Protein Microcompartment Generation By Expressing a Propanediol Utilisation Organelle from Geobacillus thermoglucosidasius in Bacillus Subtilis

ICBE Asia

Conference: International Conference Biomolecular Engineering ICBE

Year: 2018

Proceeding: ICBE Asia 2018: International Conference on Biomolecular Engineering

Group: General Submissions

Session: Session 3: Translational Biomolecular Engineering

Time: Tuesday, January 9, 2018 - 9:30am-9:55am

Authors:

Wade, Y., University of Bath
Daniel, R.
Leak, D. J., University of Bath

Bacterial microcompartments (BMCs) are bacterial sub-cellular proteinaceous organelles which host specific biochemical reactions in a manner which is physically separated from other cytoplasmic processes. This includes metabolosome housing the propanediol metabolic pathway where the purpose appears to be to sequester highly reactive pathway intermediates. Gaining control of assembly of BMCs and their biochemical cargo offers biotechnological opportunities, and this has yet to be achieved in thermophiles. In this study, we have established a strategy for construction of an empty organelle from *Geobacillus thermoglucosidasius* using components of the propanediol utilisation metabolosome and targeting of enzymes in its lumen in *Bacillus subtilis*,as model Gram-positive organism.

The >18kb *pdu* operon consisting of 19 genes was examined. The five genes, *pduA*, *pduBBâ €™*, *pduJ*, *pduK* and *pduN*, encoding proteins that form the shell of BMCs were dispersed with others. These genes were amplified from the *G. thermoglucosidasius* chromosome and assembled in the same order as in the locus and integrated into the *amyE* locus of the *B. subtilis* 168CA chromosome under control of a IPTG inducible promoter to give the BMCG strain. Initially, the ability to express heterologous proteins was confirmed by the appearance of a green fluorescent signal after two hours induction in the strains where the superfolder GFP (sfGFP) was fused in frame to each individual genes at the C-terminus. Cytoplasmic fluorescence intensity differed between the strains indicating an abundance of shell proteins with PduBBâ€™ and PduJ having the highest levels. Western immunoblotting of the whole-cell protein extract of fusion sfGFP to pduA and pduN, using anti-sfGFP antibody, revealed bands with molecular weights of 37 and 36kDa, respectively, in contrast to 27kDa in the control. Circular structures with clear boundaries were visualised in thin-sectioned cells of BMCG by TEM and were absent in the control strain with empty vector integration. In order to determine the protein "signal" sequences required for targetting sfGFP to the empty BMC, it was linked to the N-terminal sequences of Propanediol utilisation protein (PduP) equal to 18, 24 and 461 amino acids. The constructs were placed into the *lacA* genomic region of the BMCG strain under the xylose inducible promoter. Discrete fluorescent puncta were observed after co-expression of shell and both P24 and P461 sfGFP fused proteins. Separately, the P18 sfGFP fusion seemed to be unable to incorporate sfGFP in the BMCs, resulting in a diffuse fluorescent signal. Time-lapse imaging showed an ability to assemble a BMC after 20 minutes.

These results show for the first time the capacity to generate synthetic protein compartments derived from the shell proteins, PduABBâ€™JKN, of the thermophile, *G. thermoglucosidasius*. Based on studies with PduP, targeting of heterologous protein to the interior of the compartment requires fusion of an N-terminal sequence of at least 24 amino acids. The current *Bacillus*-based system will assist in future construction of empty compartments for targeting proteins in *Geobacillus* spp. at high temperatures.

Engineering Salmonella Effector Protein as an Inflammatory Bowel Disease Therapeutic

Conference: International Conference Biomolecular Engineering ICBE

Year: 2018

Proceeding: ICBE Asia 2018: International Conference on Biomolecular Engineering

Group: General Submissions

Session: Session 3: Translational Biomolecular Engineering

Time: Tuesday, January 9, 2018 - 9:55am-10:20am

Authors:

Champion, J. A., Georgia Institute of Technology

Bacterial pathogens have evolved effector proteins to suppress mammalian inflammatory and immunoregulatory pathways. One such protein, AvrA from *Salmonella*, is an acetyltransferase that inhibits inflammatory JNK and NF-kB signaling without inducing apoptosis. We have engineered AvrA protein nanoparticles to deliver AvrA intracellularly in the absence of the *Salmonella* secretion system. AvrA nanoparticles are internalized *in vitro* and *in vivo* into epithelial and monocytic cells. They inhibit inflammatory signaling and confer cytoprotection *in vitro*. In two murine colitis models, we observe decreased clinical and histological indices of inflammation for either preventative or therapeutic dosing via transrectal instillation. To implement AvrA nanoparticles as a clinically viable treatment for IBD with access to the entire intestinal tract and improved patient compliance, an oral formulation was designed. AvrA nanoparticles were encapsulated into microparticles using a microfluidic device. The microparticles protected their protein cargo in the gastric compartment and released the nanoparticles for intracellular delivery *in vitro* and *in vivo*. Prevention of colitis symptoms was observed following oral delivery of AvrA nanoparticles in microparticles. This work illustrates the potential therapeutic benefits of bacterial pathogen effectors when delivered outside the context of infection.

Co-Authors: Kevin Ling, Lina Herrera Estrada, Huixia Wu, Andrew S Neish, and Julie Champion

The EMBO Lecture: Coiled-coil protein origami designed nanostructures

Conference: International Conference Biomolecular Engineering ICBE

Year: 2018

Proceeding: ICBE Asia 2018: International Conference on Biomolecular Engineering

Group: General Submissions

Session: The EMBO Keynote Lecture

Time: Tuesday, January 9, 2018 - 1:30pm-2:05pm

Authors:

Jerala, R., National Institute of Chemistry

Proteins are the most advanced nanostructures, defined by the sequence of amino acids. Nature provides a limited number of protein sequences and folds, which have been optimized during evolution. Modular design of polypeptide-based polyhedra was inspired by the DNA-based nanostructures. The folding problem of proteins was bypassed by designing *de novo* **coiled-coil protein origami (CCPO) folds** based on the well-understood specificity of coiled-coil dimers and used them as modules to guide the polypeptide chain between vertices of the selected polyhedral cage. The principle was demonstrated by the construction of a nanoscale protein tetrahedral cage from a single polypeptide chain composed of 12 coiled-coil forming segments[1] . In this assembly 6 edges of the polyhedron are defined by orthogonal coiled-coil dimers. Introduction of supercharged CC building modules[2] and protein origami computational design platform CoCoPOD enabled design of protein origami cages that are able to self-assemble in vivo. Complexity of protein origami structures as extended to more than 700 residue proteins with tetrahedral, four-sided pyramid and triangular prism that fold efficiently, with kinetics and stability comparable to globular proteins[3] . *In vivo*folding of protein origami opens prospects for new applications of this new type of designed nanostructures.

- GradiÅ¡ar H, BoÅ¾iÄ S, Doles T, Vengust D, Hafner-BratkoviÄ I, Mertelj A, Webb B, Å ali A, KlavÅ¾ar S, Jerala R. (2013) ***Nat Chem Biol***. 6:362-6.
- Drobnak I, GradiÅ¡ar H, LjubetiÄ A, Merljak E, Jerala, R. (**2017**) Modulation of Coiled-Coil Dimer Stability through Surface Residues while Preserving Pairing Specificity. ***Am.Chem.Soc.***139: 8229-8236.
- LjubetiÄ, A., Lapenta, F., GradiÅ¡ar, H., Drobnak, I., AupiÄ, J., StrmÅ¡ek, Å½., LainÅ¡Äek, D., Hafner-BratkoviÄ, I., Majerle, A., Krivec, N., BenÄina, M., Pisanski, T., Ä†irkoviÄ‡ VeliÄkoviÄ‡, T., Round, A., Carazo, J.M., Melero, R. and Jerala, R., (**2017**) Design of in vivo self-assembling coiled-coil protein origami. ***Nat Biotech***, 35:1094-1101.

Invited Talk: Repurposing Nature: Engineering Protein Nanocage for Medicine and Beyond

Conference: International Conference Biomolecular Engineering ICBE

Year: 2018

Proceeding: ICBE Asia 2018: International Conference on Biomolecular Engineering

Group: General Submissions

Session: Session 4: Biomolecular Programming - from DNA to Community

Time: Tuesday, January 9, 2018 - 2:05pm-2:40pm

Authors:

Lim, S., Nanyang Technological University

Nature provides us with abundance of biological materials and we have been repurposing them to suit our needs. In this talk, I will highlight protein nanocages and our attempts to engineer, reformat, and repurpose them for medicine, molecular electronic, and consumer care applications.

Protein nanocages can be engineered to tailor their functions. They are formed by the self-assembly of multiple subunits forming hollow cage-like structures of nanometer size. Due to their proteinaceous nature, the protein nanocages allow facile modifications on its internal and external surfaces, as well as the subunit interfaces designed for the intended applications. Protein nanocages loaded with metal have properties that are promising as MRI contrast agents. Loading them with drugs and modifying the interface of the subunits to render the nanocages sensitive to environmental changes, such as pH, shows their potential application as drug carriers.Engineering of the external surface allows for the display of targeting ligands for selective accumulation on cancer cells. Leveraging on its natural or engineered metal-chelating activities, protein nanocages serve a dual function as a reaction container and as a facilitator in the deposition of monodispersed platinum nanoparticles on graphene surfaces for electrocatalysis in fuel cells. Long-range electron tunneling across metal-loaded protein nanocages has also been shown to be promising in the development of memristive devices and future molecular electronics. In the most recent works, we show that the protein nanocages are surface active with an ability to stabilize Pickering emulsion with pH-responsive behavior.

An Electronic Analogous Synthetic Genetic 2-to-4 Digital Decoder and Associated Molecular Tools in Living Cell

ICBE Asia

Conference: International Conference Biomolecular Engineering ICBE

Year: 2018

Proceeding: ICBE Asia 2018: International Conference on Biomolecular Engineering

Group: General Submissions

Session: Session 4: Biomolecular Programming - from DNA to Community

Time: Tuesday, January 9, 2018 - 2:40pm-3:05pm

Authors:

Bagh, S., Massachusetts Institute of Technology

Adapting engineering design principle in the realm of molecular cell biology, synthetic biologists developed few artificial cellular information-processing systems in recent years. Those systems can process extracellular chemical signal to count, to communicate and to show different patterns in response. One of the further challenges is to develop novel system that can process multiple extracellular chemical signals together and decode different combination of signals to make distinct decision for each and every possible combination. Here we demonstrate our recent effort in creating a population level 2-to-4 genetic decoder using synthetic gene circuits in living *Escherichia coli* by following digital electronics principle. A 2-to-4 decoder has two input signals and based on the combination of inputs, the device can take four distinct decisions. Here we have designed and constructed a set of synthetic genetic AND, NOR, AND (NOT) and (NOT) AND gates by creating a library of novel hybrid promoters and novel truncated proteins with non-conventional start codon and characterized them experimentally. The best candidates with digital behavior have been taken for full circuit assembly. However, assembling several genetic cascades in a single plasmid is a challenge. Here we developed a modified BIOBRICK assembly technique, where multiple genetic cascades can be assembled in a single plasmid in any direction, a characteristic lacked in previous systems. In our system we have two extra cellular chemicals as input signals and 4 different fluorescent proteins as distinct outputs. Such systems have potential application in artificial cellular communication systems, cellular robotics and programmed therapeutics.

Controlling Bdellovibrio Bacteriovorus Gene Expression and Predation Using Synthetic Riboswitches

Conference: International Conference Biomolecular Engineering ICBE

Year: 2018

Proceeding: ICBE Asia 2018: International Conference on Biomolecular Engineering

Group: General Submissions

Session: Session 4: Biomolecular Programming - from DNA to Community

Time: Tuesday, January 9, 2018 - 3:05pm-3:30pm

Authors:

Dwidar, M., Okinawa Institute of Science and Technology
Yokobayashi, Y., Okinawa Institute of Science and Technology

Bdellovibrio bacteriovorus is a predatory bacterium that feeds on Gram-negative bacteria including a wide range of pathogens and thus has potential applications as a biocontrol agent. Owing to its unique life cycle, however, there are limited tools that enable genetic manipulation of *B. bacteriovorus*. This work describes our first steps towards engineering the predatory bacterium for practical applications by developing basic genetic parts to control gene expression. Specifically, we evaluated four robust promoters that are active during the attack phase of *B. bacteriovorus*. Subsequently, we tested several synthetic riboswitches that have been reported to function in *Escherichia coli*, and identified theophylline-activated riboswitches that function in *B. bacteriovorus*. Finally, we inserted the riboswitch into the bacterial chromosome to regulate expression of the flagellar sigma factor *fliA* which was previously predicted to be essential for predation, and observed that the engineered strain shows a faster predation kinetics in the presence of theophylline.

Biosynthesis of Triacsins

ICBE Asia

Conference: International Conference Biomolecular Engineering ICBE

Year: 2018

Proceeding: ICBE Asia 2018: International Conference on Biomolecular Engineering

Group: General Submissions

Session: Session 4: Biomolecular Programming - from DNA to Community

Time: Tuesday, January 9, 2018 - 3:30pm-3:55pm

Authors:

Twigg, F., UC Berkeley

Nitrogen-nitrogen (N-N) bonds are found in a variety of natural products with diverse biological activities, but only two enzymatic pathways resulting in N-N bond formation have been characterized. The triacsin family contains four known congeners isolated from *Streptomyces aureofaciens*, and is commonly used today for its potent inhibition of acyl-CoA synthetases. Triacsins are notable for their *N*-hydroxylated triazene moiety, which is not found in any other known natural products. We have recently identified the biosynthetic gene cluster for triacsins in*Streptomyces aureofaciens*, and further genome mining effort led to the discovery of an additional triacsin native producer. Our *in vivo* and *in vitro* analysis also revealed new biosynthetic insights regarding triacsin biosynthesis in *Streptomyces*.

A Method for High-Throughput Screening of Initial Codons Engineered for Maximal Production of Recombinant Proteins

Conference: International Conference Biomolecular Engineering ICBE

Year: 2018

Proceeding: ICBE Asia 2018: International Conference on Biomolecular Engineering

Group: Poster Session

Session: Poster Session

Time: Tuesday, January 9, 2018 - 4:05pm-5:30pm

Authors:

Park, Y. J., Chungnam National university
Lee, K. H., Chungnam National University
Kim, D. M., Chungnam National University

Cell-free protein synthesis system provides an open environment for gene expression and allows direct access to the molecular events of transcription and translation. In this study, we demonstrate the use of cell-free protein synthesis techniques for determining the initial codons that can lead to maximized production of target proteins. We first randomized the first two codons, and the individual variant genes were co-expressed with the gene of a reporter protein (sfGFP) in the same mixture. As a results of the competition between the variant gene and a common reporter gene for the finite pool of ribosome in the reaction mixture, we could readily estimate the relative expression levels of the variant genes by inverse correlation between the sfGFP fluorescence and the expression levels of target genes [3]. Through this simple method, we could rapidly screen for the initial codons that substantially enhances the production of a target protein. Importantly, the initial codons identified by the cell-free co-expression method also effectively enhanced protein production in *E. coli,* indicating that this method can be used as a general tool to predict and modulate the yield of protein production during cell-based gene expression.

Bicontinuous Interfacially Jammed Emulsion Gels for Efficient Enzyme-Catalyzed Conversion of Poorly Water-Soluble Substrate

ICBE Asia

Conference: International Conference Biomolecular Engineering ICBE

Year: 2018

Proceeding: ICBE Asia 2018: International Conference on Biomolecular Engineering

Group: Poster Session

Session: Poster Session

Time: Tuesday, January 9, 2018 - 4:05pm-5:30pm

Authors:

Cha, S.
Lim, H. G., POSTECH
F Haase, M., Rowan University
Jung, G. Y., POSTECH
Lee, D., University of Pennsylvania

Enzyme is biocatalyst which has been widely used in bio-industrial processes due to its several advantageous characteristics. However, platform technology for the the enzymatic reaction containing water-insoluble substrate remains undeveloped, because most of the enzymes are active in only aqueous phase. In this study, to expedite the enzymatic reaction, we utilized bicontinuous interfacially jammed emulsion gel (bijel) which has enormous interfacial area between two immiscible liquids.

Using tributyrin hydrolysis as a model reaction, we provide evidence that the interfacial area between two immiscible liquids of bijel is a crucial factor for the efficient enzymatic reaction involving poorly water-soluble substrate. Moreover, bijel system showed the 3.91 fold increase in reaction rate than that of biphasic system which is the most common method for the enzymatic conversion of oil-soluble substance. The outcomes indicate that bijel could be a platform material to accelerate the conversion of various hydrophobic materials.

Formatting Issue - Best Available Paper/Graphic

Cell Surface Display of Mussel Inspired Catecholamine - the Application of Sticky Bacteria

ICBE Asia

Conference: International Conference Biomolecular Engineering ICBE

Year: 2018

Proceeding: ICBE Asia 2018: International Conference on Biomolecular Engineering

Group: Poster Session

Session: Poster Session

Time: Tuesday, January 9, 2018 - 4:05pm-5:30pm

Authors:

Lee, S. H., Chonnam National University
Park, C. W., Chonnam National University

Cell surface display is a technique to display peptides or proteins on the surface of gram-negative and gram-positive bacteria, fungi, or even mammalian cells by appropriately fusing them to surface anchoring motifs. [1,2]

Here, we present a sticky *Escherichia coli* which displayed catecholamine, a mussel-inspired adhesive moiety with catechol and amine functional group, on the cell surface using outer membrane protein W of *E. coli* as an anchoring motif. The catecholamine displayed cells exhibited rapid cohesion between cells to form cellular aggregates and gold nanoparticles were reduced on the cell surface when incubated with gold solution.[3] As an application, enantioselective resolution, robust adhesion between the engineered cells and various substrates, and adsorption of heavy metals were observed. Sticky *E. coli* displaying mussel-inspired catecholamine reported here can be a versatile platform microbe for various emerging fields such as whole living cell attachment on versatile material surfaces, cell-to-cell communication systems, and many others.

[1] Lee, S. Y.; Choi, J. H.; Xu Z., J. *Trends Biotechnol.* **2003**, *21*, 45Ë—52.

[2] Han, M. J., Lee, S. H., *FEMS Microbiol. Lett.* **2014**, *362*, 1Ë—7.

[3] Park, J. P., Choi, M. J., Kim, S. H., Lee, S. H., Lee, H., *Appl. Environ. Microbiol.* **2014**, *80*, 43Ë—53.

Characterization of Recombinant Saccharomyces Cerevisiae Expressing a Mutated SPT15p Global Transcription Factor in Microaerobic Fermentations

ICBE Asia

Conference: International Conference Biomolecular Engineering ICBE

Year: 2018

Proceeding: ICBE Asia 2018: International Conference on Biomolecular Engineering

Group: Poster Session

Session: Poster Session

Time: Tuesday, January 9, 2018 - 4:05pm-5:30pm

Authors:

Park, Y. C.
Park, H. S.
Seong, Y. J.

The *SPT15* gene encodes a *Saccharomyces cerevisiae* TATA-binding protein, which is able to globally control the transcription levels of various metabolic and regulatory genes. In this study, a *SPT15* gene mutant (S42N, S78R, S163P and I212N) was expressed in *S. cerevisiae* BY4741 (SPT15-M3), of which effects on the yeast cell properties were evaluated in batch, fed-batch and simultaneous saccharification and fermentation (SSF) processes. Organic nitrogen sources and a microaerobic condition were more favorable for SPT15-M3 than the SPT15wt control in both cell growth and ethanol production. Fed-batch cultures of SPT15-M3 using concentrated glucose solution resulted in 9-19 % higher glucose consumption rate and ethanol productivity than those for SPT15wt. In addition, SPT15-M3 showed 3.9 and 4.5% increases in ethanol productivity from cassava hydrolysates and corn starch in SSF processes, respectively. It was concluded that overexpression of mutated *SPT15*gene in *S. cerevisiae* would be a potent strategy to enhance ethanol production from glucose-based biomass.

Complementary Cell-Free Protein Synthesis Assay for Quantification of Amino Acids

Conference: International Conference Biomolecular Engineering ICBE

Year: 2018

Proceeding: ICBE Asia 2018: International Conference on Biomolecular Engineering

Group: Poster Session

Session: Poster Session

Time: Tuesday, January 9, 2018 - 4:05pm-5:30pm

Authors:

Jang, Y. J., Chungnam National University
Lee, K. H., Chungnam National University
Kim, D. M., Chungnam National University

The concentration of amino acids in blood is often related to different diseases directly or indirectly. In this study, we present a simple and economical method that enables rapid quantification of amino acids based on their polymerization into a signal-generating fluorescence protein. In this method that harnesses an amino acid-deficient cell-free synthesis system as a conditional signal-generating module, fluorescence signal is generated in response to the presence of specific amino acids in assay samples. When premixed with assay samples containing the amino acids in question, incubation of the cell-free synthesis reaction mixture rapidly resulted in the production of sfGFP, the fluorescence intensity of which was linearly proportional to the concentration of the amino acids. The assay method achieved a limit of detection as low as ~100 nM, and was successfully applied to the quantification of disease-related amino acids in biological samples. Compared to existing methods relying on HPLC, the complementation assay method developed in this work enables the direct translation of amino acid titer into measurable bio-fluorescence intensity in a much shorter period, providing a more affordable and flexible option for the quantification of amino acids.

Development of High Efficient Systems for Producing Copper Peptide As a Cosmetic Material in the Yeast

ICBE Asia

Conference: International Conference Biomolecular Engineering ICBE

Year: 2018

Proceeding: ICBE Asia 2018: International Conference on Biomolecular Engineering

Group: Poster Session

Session: Poster Session

Time: Tuesday, January 9, 2018 - 4:05pm-5:30pm

Authors:

Yu, B. J., KITECH
Jang, J. Y., KITECH
Park, S., KITECH(Korea Institute of industrial Technology)
Choi, S., KITECH

Copper peptide is composed of three amino acids (Gly-His-Lys) and copper and naturally in human plasma. Since it was discovered in 1973, it has been reported that copper peptide plays wide roles in activation of wound healing, attraction of immune cells, antioxidant and anti-inflammatory effects, and stimulation of collagen synthesis. Especially, copper peptide is widely used in anti-aging cosmetics. However, copper peptide has been produced by using chemical synthesis. In this study, we developed a biological system to produce and purify this three peptide (Gly-His-Lys) in a GRAS yeast, Pichia pastoris. First, we constructed an expression vector (pBJY-GHK6) in which six of Gly-His-Lys fused to His-tag could be expressed and secreted into media broth. The linearized pBJY-GHK6 was transformed and integrated into genome of P. pastorisand the recombinant strain was cultivated in minimal media (YNB/YSD without adenine). We used an affinity column (Ni-column) to purify the expressed cassettes, [$(His)_6$-K-$(GHK)_6$] and [$(GHK)_6(His)_6$] and treated the expressed peptide with trypsin to generate GHK monomers. We confirmed the expressed peptide on SDS-PAGE and analyzed the GHK monomers by using HPLC and ESI-MS. We are developing optimized condition to increase concentration of the expressed GHK by optimizing culture media, fermentation mode and purification process. In addition, trypsin is going to also be produced in Pichia pastoris and then used in the cleavage of [$(His)_6$-K-$(GHK)_6$] and [$(GHK)_6(His)_6$]. This system will be applied to widely be used in production of cosmetic, cosmeceutical and even pharmaceutical peptides.

Direct Conversion of CO2 to Squaleneby Metabolically-Engineered in Cyanobacteria

Conference: International Conference Biomolecular Engineering ICBE

Year: 2018

Proceeding: ICBE Asia 2018: International Conference on Biomolecular Engineering

Group: Poster Session

Session: Poster Session

Time: Tuesday, January 9, 2018 - 4:05pm-5:30pm

Authors:

Choi, S. Y., Sungkyunkwan University
Sim, S. J., Korea University
Woo, H. M., Sungkyunkwan University

Direct conversion of CO_2 to squalene

by metabolically-engineered in cyanobacteria

Sun Young Choi[1], Sang Jun Sim[2], Han Min Woo[1]*

[1] Department of Food Science and Biotechnology, Sungkyunkwan University, 2066, Seobu-ro, Jangan-gu, Suwon, Republic of Korea

[2] Department of Chemical and Biological Engineering, Korea

University, 145, Anam-ro, Seongbuk-gu, Seoul, Republic of Korea

*Correspondence
to hmwoo@skku.edu, TEL: +82-31-290-7808, FAX: +82-31-290-7882

Metabolic
engineering of cyanobacteria has enabled photosynthetic conversion of CO_2
to value added chemicals as bio-solar cell factories [1]. The
using cyanobacterial for direct production of squalene is advantage, since they
can just grow carbon dioxide from air and sunlight without energy sources based
on glucose.
However, the production levels of squalene among several
isoprenoids in engineered cyanobacteria were quite low, compared to other
microbial hosts. Here, we engineered S.
elongatus PCC 7942 with modular metabolic pathways consisting of
the methylerythritol phosphate pathway enzymes and the squalene synthase for
production of squalene, originated from 4 type microorganisms [2]. Sequentially, to increase
production of squalene
from CO_2 by application ofthe push-and-pull
strategy. Squalene synthase
(SQS) was fused to either a key enzyme (farnesyl diphosphate synthase) of the
methylerythritol phosphate pathway or the ¥â-subunit of phycocyanin(CpcB1).
Moreover, the best squalene producer was cultivated in a scalable
photobioreactor (6 L) with light optimization [3]. Further
development of the photo-bioprocessing conditions with strain improvement will
promote establishment
of an engineered bio-solar cell factory for industrial-scale CO_2
conversion. This work was financially supported by Korea CCS 2020 Program
granted by National Research Foundation of Korea.

Keywords: Metabolic
engineering, Cyanobacteria, Synthetic biology, Isoprenoids, squalene

Reference

[1] Woo, H.M., Solar-to-chemical
and solar-to-fuel production from CO_2 by metabolically engineered
microorganisms. Curr. Opin. Biotechnol.
2017, 45:1–7.

[2] Choi, S.Y.; Lee, H.J.; Choi,
J.; Kim, J.; Sim, S.J.; Um, Y.; Kim, Y.; Lee, T.S.; Keasling, J.; Woo, H.M.,

Photosynthetic conversion of CO_2 to farnesyl diphosphate-derived
phytochemicals (amorpha-4,11-diene and squalene) by engineered cyanobacteria, Biotechnol. Biofuels. 2016, 9:202.

[3] Choi, S.Y., Wang, J.Y., Kwak,
H.S., Lee, S.M., Um, Y., Kim, Y., Sim, S.J., Choi. J.-I., *Woo, H.M.,
Improvement of squalene production from CO2 in Synechococcus elongatus PCC 7942
by metabolic engineering and scalable production in a photobioreactor (2017.07)
ACS Synth. Biol. 6(7) p. 1289-1295

Efficient Production of D-Lactate from Methane Using Lactate-Tolerant Methylomonas Sp. DH-1 Strain Generated By Adaptive Laboratory Evolution

Conference: International Conference Biomolecular Engineering ICBE

Year: 2018

Proceeding: ICBE Asia 2018: International Conference on Biomolecular Engineering

Group: Poster Session

Session: Poster Session

Time: Tuesday, January 9, 2018 - 4:05pm-5:30pm

Authors:

Kim, W., Seoul National University
Hahn, J. S., Seoul National University
Kim, S., Seoul National University
Lee, J., Seoul National University

Metabolic engineering of methanotrophs, which can use methane as a sole carbon and energy source, has emerged as a promising strategy to utilize methane, an abundant and inexpensive carbon source. In this study, we engineered *Methylomonas* sp. DH-1 to produced D-lactate from methane. *Methylomonas* sp. DH-1 belongs to type I methanotrophs, which assimilate methane by ribulose monophosphate (RuMP) pathway. Wild type strain can survive only up to 1.0 g/L lactate. Therefore, we improved lactate tolerance of *Methylomonas* sp. DH-1 by growing the cells with increasing concentrations of lactate up to 8.0 g/L, generating a lactate-tolerant strain Ev 8.0. D-Lactate production was achieved by integrating D form-specific lactate dehydrogenase gene (*ldhA*) from*Leuconostoc mesenteroides* subsp. *mesenteroides* ATCC 8293 into the genome of Ev 8.0. In a flask fed-batch culture with pH and medium optimization, the engineered strain produced 1.36 g/L D-lactate from methane after 144 h.

Electrochemical Detection of Radical Scavenging Polyphenols in Natural Honey: Structure-Activity Relationships

Conference: International Conference Biomolecular Engineering ICBE

Year: 2018

Proceeding: ICBE Asia 2018: International Conference on Biomolecular Engineering

Group: Poster Session

Session: Poster Session

Time: Tuesday, January 9, 2018 - 4:05pm-5:30pm

Authors:

Ismail, N. I., Universiti Teknologi Malaysia
Sundaram, S.
Abdul Kadir, M. R., Universiti Teknologi Malaysia
Mohamed Zulkifli, R.
Shahir, S.

Consumption of antioxidant-rich foods is very likely helpful in protecting human body from deleterious effects of free-radical induced oxidative stress. Polyphenols, the plant secondary metabolites that have been recognized to serve as antioxidants, are among hundreds of minor constituents present in natural honeys. The present work aims to analyze the electrochemical behaviour of polyphenols identified in Malaysian honey samples and elucidate their structure-radical scavenging activity relationships. A total of 18 polyphenols comprises of nine flavonoids, eight phenolic acids, and tannin as identified from our earlier work was studied. Oxidation-reduction (redox) properties of these polyphenols prepared in 0.1 M PBS at pH 7 were analyzed using cyclic voltammetry (CV). The results of the study as shown by cyclic voltammograms indicated that polyphenols with catechol moiety specifically at the B-ring exhibited first anodic (oxidation) peak at low potentials, ranging between +0.173V and +0.404V. Subsequent anodic peaks appeared at much higher potentials were seen for polyphenols bearing a third phenol group (gallic moiety) or other electroactive group (resorcinol moiety). For polyphenols with only isolated phenol group on the B-ring, anodic peak(s) were observed at higher potentials as well. The presence of cathodic (reduction) peaks was observed in several analysed phenolics. It is concluded that the polyphenols varied in their electron donor ability as influenced by their structure as well as hydroxyl groupâ€™s number and location, suggesting that they exhibit different radical scavenging activity.

Engineering an Aldehyde Dehydrogenase with Respect to Its 2 Substrates for Improving 3-Hydroxypropionic Acid Production

Conference: International Conference Biomolecular Engineering ICBE

Year: 2018

Proceeding: ICBE Asia 2018: International Conference on Biomolecular Engineering

Group: Poster Session

Session: Poster Session

Time: Tuesday, January 9, 2018 - 4:05pm-5:30pm

Authors:

Park, Y. S., Ajou university
Choi, S. J., Ajou university
Yoo, T. H., Ajou University
Nguyen, N. H., Ulsan National Institute of Science and Technology
Park, S., UNIST
Choi, U. J., Ajou University

3-Hydroxypropionic acid (3-HP) can be produced via two enzymatic reactions: dehydration of glycerol to 3-hydroxypropanal (3-HPA) and oxidation to 3-HP. Commercial production of 3-HP has been beset with several problems. Some of these problems are associated with the toxicity of 3-HPA and the efficiency of NAD^+ regeneration. We engineered Î±-ketoglutaric semialdehyde dehydrogenase (KGSADH) for the second reaction to address these issues. The residues in the putative binding sites for the substrates, 3-HPA and NAD^+, were randomized, and the libraries were screened for higher activity. Isolated KGSADH variants had lower Km values for both substrates. The enzymes showed higher substrate specificities for aldehyde and NAD^+, less inhibition by NADH, and greater resistance to inactivation by 3-HPA than the wild-type enzyme. A recombinant *Pseudomonas denitrificans* strain with one of the engineered KGSADH variants exhibited less accumulation of 3-HPA, decreased levels of inactivation of the enzymes involved in the production of 3-HP. These attributes facilitated sustained production of 3-HP in to the late stages of culture and enhanced the final titer of 3-HP by approximately 40%.

Expression and in Vitro Reconstitution of Novel Lasso Peptide Gene Clusters

Conference: International Conference Biomolecular Engineering ICBE

Year: 2018

Proceeding: ICBE Asia 2018: International Conference on Biomolecular Engineering

Group: Poster Session

Session: Poster Session

Time: Tuesday, January 9, 2018 - 4:05pm-5:30pm

Authors:

Koos, J.

Lasso peptides are members of a class of small molecules called ribosomally synthesized post-translationally modified peptides (or RiPPs) characterized by a unique three-dimensional structure. Lasso peptides are defined by a single isopeptide bond formed between the N-terminus and an acidic side chain, forming a ring. The remaining C-terminus of the peptide is threaded through this ring, forming the characteristic lariat shape. Recent research has shown that potential lasso peptide clusters are widespread in a range of genomes, but the set of isolated lasso peptides is more limited. A set of gene clusters were found through genome mining with peptides predicted as more hydrophobic than known examples and with previously unobserved N-terminal residues. These clusters were heterologously expressed and the properties of the products were investigated. In addition, the associated maturation proteins for each cluster were expressed and purified. Using purified components, in vitro experiments were performed and mature peptides were identified by LC-MS, confirming the activity of the proteins. The in vitro reconstituted system also allowed quantification of the activity of the enzymes in a manner that was not possible through heterologous expression of clusters and enables future studies to further understand the details of lasso peptide production.

Genome Sequence of Potential Eukaryotic Probiotics Saccharomyces Cerevisiae Rey 36-3 Producing Glutathione

ICBE Asia

Conference: International Conference Biomolecular Engineering ICBE

Year: 2018

Proceeding: ICBE Asia 2018: International Conference on Biomolecular Engineering

Group: Poster Session

Session: Poster Session

Time: Tuesday, January 9, 2018 - 4:05pm-5:30pm

Authors:

Kim, H. J., Seoul National University

Nuruk is a starter for the Korean rice wine *Makgeolli.* It consists of various microbes such as yeasts, fungi, some lactic acid bacteria. Since*Makgeolli* has been consumed by Korean for a long time, yeast collected from *Nuruk* can be safely used as probiotics. In addition, glutathione produced by yeast cells may boost the adaptive immune system and scavenging the intestinal oxidative environment. Two hundred seventy wild yeast strains were collected from *Nuruk* and the amounts of glutathione in the yeast cells were estimated. The wild *S. cerevisiae* Rey 36-3 strain was selected among 270 yeast strains for the glutathione production and probiotic uses. We sequenced the *S. cerevisiae* Rey 36-3 strain using PacBio sequencing platform and obtained the 12-MB near complete genome sequence containing 5,665 predicted genes. The genome annotation revealed 21 genes involved in the glutathione synthesis pathway. The whole genome sequence of the yeast strain will provide useful information to engineer the strain for producing glutathione.

Metabolic Engineering of Fatty Acid Downstream Pathway in Saccharomyces Cerevisiae to Produce Olechemicals

ICBE Asia

Conference: International Conference Biomolecular Engineering ICBE

Year: 2018

Proceeding: ICBE Asia 2018: International Conference on Biomolecular Engineering

Group: Poster Session

Session: Poster Session

Time: Tuesday, January 9, 2018 - 4:05pm-5:30pm

Authors:

Zhang, X., Beijing University of Chemical Technology

With the rapid development of synthetic biology and metabolic engineering, microbial production of biofuels has become an attractive alternative instead of traditional fossil fuels. The budding yeast*Saccharomyces cerevisiae* as a robust model microorganism is widely used in metabolic engineering. Here, we engineered downstream pathway of fatty acid to produce fatty acid-derived chemicals in *S. cerevisiae*. First, we constructed a more effective downstream pathway to directly converted FFAs to alkene by an appropriate P450 and overexpressed *ACC1* gene (acetyl-CoA carboxylase) simultaneously to increase the production of malonyl-CoA. Besides, we also designed another self-sufficient P450 to produce terminal hydroxyl fatty acid by FFAs and could be further converted to amino fatty acid which is a monomer for an important synthetic fibre-nylon. At last, FFAs were firstly reduced to fatty aldehyde and then further converted to alkane by ADO. Fusion expression of ADO with alternative electron transfer system could efficiently converted fatty aldehyde to alkane in *S. cerevisiae*. The production of fatty acid-derived chemicals (alkene, hydroxyl fatty acid and alkane) was significantly improved by the three optimized pathways in *S. cerevisiae* that provides an efficient and economic route to the important target chemicals.

Microbial sequencing in Singapore, a journey from the wet lab to the dry lab

ICBE Asia

Conference: International Conference Biomolecular Engineering ICBE

Year: 2018

Proceeding: ICBE Asia 2018: International Conference on Biomolecular Engineering

Group: Poster Session

Session: Poster Session

Time: Tuesday, January 9, 2018 - 4:05pm-5:30pm

Authors:

Florez de Sessions, P.

As humans, we share our existence with a complex community of microbial partners; fungal, viral and bacterial. At the Genome Institute of Singapore (GIS), I run a platform, the **G**IS **E**fficient **R**apid **M**icrobial**S**equencing (GERMS) platform, which utilizes both wet lab and dry lab tools to genetically characterize these populations and how they impact our lives. The principal goal of the platform is to enable industrial and academic partners to use microbial genomics for a variety of applications. We have used our tools to help medicinal chemists with data mining after rational drug design with Novartis (1), used comparative genomics and transcriptomics to decipher novel gene families that can explain how that fungi can survive on lipids from the skin, and potentially serve as therapeutic targets with P&G (2).

We also collaborate with academics in South East Asia such as the Oxford Clinical Research Unit in Vietnam and Duke-NUS in Singapore to understand microbial causes for the infectious diseases prominent in this part of the world. In August 2016 Zika was simultaneously reported by various sentinel surveillance sites around Singapore and the Ministry of Health (MOH) declared a public health emergency. In an island wide effort involving the MOH, Singaporean hospitals, Duke-NUS, GIS and the National Environmental Agency (NEA), we managed to rapidly sequence, analyze and curb the spread of infection. The outbreak tracking of this pathogen was conducted in both mosquitos and humans (3). Our whole genome sequences were deposited in the NCBI database and doubled the existing whole genome sequences for Zika in the database at the time of publication. In this study we describe how our investigation and control of the Zika outbreak in Singapore, by detailing the effort to contain and control vector mosquito populations by the NEA, which likely contributed to a reduction in disease transmission within weeks of discovery.

The world of genomics and the data mining process is an ever changing and challenging landscape. However, if we stay on top of the innovative wet lab methods and current bioinformatics tools available the power of informing public health through genomics is vast.

References:

1. Exploring the mode of action of bioactive compounds by microfluidictranscriptional profiling in mycobacteria. PLoS One, PMID 23935951.

2. Genus-wide comparative genomics of Malassezia delineates its phylogeny, physiology, and niche adaptation on human skin. PLoS Genetetics, PMID 26539826.
3. Outbreak of Zika in Singapore â€" a Clinical, Epidemiological, Phylogenetic , and Entomological Account. Lancet Infectious Disease, PMID 28527892

Multi-Omics Analysis of Methanotroph

ICBE Asia

Conference: International Conference Biomolecular Engineering ICBE

Year: 2018

Proceeding: ICBE Asia 2018: International Conference on Biomolecular Engineering

Group: Poster Session

Session: Poster Session

Time: Tuesday, January 9, 2018 - 4:05pm-5:30pm

Authors:

Han, Y. H., Seoul National University

Methanotrophic bacterium can utilize methane, one of the next-generation chemical feedstocks, as a sole carbon source and energy to produce a variety of biofuels and chemical precursors. Recently, a novel methanotroph, *Methylomonas* sp. DH-1, was discovered and thought to be a promising strain as it has many strong points: relatively rapid growth than other methanotrophs, well-developed isoprenoid pathway, high ethanol tolerance, and so on. However, genetic information of this strain is not clear, which makes it difficult to take advantage of it in industries. To solve this problem, multi-omics analysis of this strain is required. In this study, we made the protocols for making various omics libraries specific to methanotrophs. Many methanotrophs have intracytoplasmic membrane where methane monooxygenase is attached. They also produce a variety of second metabolites and effuse them, which may take on a role of cushion when cells are lysed. As a result, cell lysis and total transcript extraction step is the major bottleneck of making libraries of methanotrophs. Therefore, we firstly optimized cell lysis method, and then developed protocols for RNA-seq, TSS-seq, and ribosome profiling specifically tailored to methanotrophs. We also made libraries of *Methylomonas* sp. DH-1 and *Methylosinus trichosporium*OB3b. By analyzing multi-omics data, we expect to understand the genetic information and metabolic pathways of methanotrophs well and make use of them to produce valuable products.

Production of Itaconic Acid from Acetate By Engineering Escherichia coli

Conference: International Conference Biomolecular Engineering ICBE

Year: 2018

Proceeding: ICBE Asia 2018: International Conference on Biomolecular Engineering

Group: Poster Session

Session: Poster Session

Time: Tuesday, January 9, 2018 - 4:05pm-5:30pm

Authors:

Woo, S. H., POSTECH
Noh, M. H., POSTECH
Lim, H. G., POSTECH
Jung, G. Y., POSTECH

Utilization of abundant and cheap carbon sources can effectively reduce the production cost and enhance the economic feasibility. Acetate is a promising carbon source to achieve cost-effective microbial processes. In this study, we engineered an *Escherichia coli* strain to produce itaconic acid from acetate. Since acetate is known to inhibit cell growth, we initially screened for a strain with a high tolerance to 10 g/L of acetate in the medium, and the W strain was selected as the host. Subsequently, itaconic acid production strain was obtained by overexpression of *cad*(encoding *cis*-aconitate decarboxylase) using a synthetic promoter and 5â€² UTR. However, the strain produced only 0.13 g/L itaconic acid because of low acetate uptake. To improve the production, the acetate assimilating pathway and glyoxylate shunt pathway were amplified by overexpression of pathway genes as well as its deregulation. The resulting strain produced 3.57 g/L itaconic acid (16.1% of theoretical maximum yield) after 88 h of fermentation with rapid acetate assimilation. These efforts support that acetate can be a potential feedstock for biochemical production with engineered *E. coli*.

Proteomic Analysis of Chlamydomonas Reinhardtii Mutant with Enhanced Lipid Production

Conference: International Conference Biomolecular Engineering ICBE

Year: 2018

Proceeding: ICBE Asia 2018: International Conference on Biomolecular Engineering

Group: Poster Session

Session: Poster Session

Time: Tuesday, January 9, 2018 - 4:05pm-5:30pm

Authors:

CHOI, J. I., Chonnam National University
Park, S., Chonnam National University

The Chlamydomonas reinhardtii mutant accumulating 20% more lipid than the wild-type, was selected. Thin-layer chromatography revealed the triglyceride and free fatty acid contents to be markedly increased in mutant strain. The major fatty acids identified were palmitic acid, oleic acid, linoleic acid and linolenic acid. Random amplified polymeric DNA analysis showed partial genetic modifications in mutant. To ascertain the changes of protein expression in the mutant strain, two-dimensional electrophoresis was conducted.

Proteomic Analysis of Scenedesmus Dimorphus Mutant with Higher Lipid Content

Conference: International Conference Biomolecular Engineering ICBE

Year: 2018

Proceeding: ICBE Asia 2018: International Conference on Biomolecular Engineering

Group: Poster Session

Session: Poster Session

Time: Tuesday, January 9, 2018 - 4:05pm-5:30pm

Authors:

CHOI, J. I., Chonnam National University

A mutant strain of the microalgae Scenedesmus dimorphus showed morphological changes and had higher growth rate compared to the wild type. In comparisons of lipid content between wild type and mutant using thin-layer chromatography, the content of triacylglycerol was markedly higher in the mutant strain. To define the metabolic changes in the mutant strain, 2-dimensional electrophoresis was conducted. Several proteins related to lipid synthesis and energy metabolisms were overexpressed in the mutant strain.

Stimuli-Responsive Protein Engineered Block Polymer Hydrogels

Conference: International Conference Biomolecular Engineering ICBE

Year: 2018

Proceeding: ICBE Asia 2018: International Conference on Biomolecular Engineering

Group: Poster Session

Session: Poster Session

Time: Tuesday, January 9, 2018 - 4:05pm-5:30pm

Authors:

Olsen, A., New York University
Montclare, J. K., NYU
Katyal, P.
Haghpanah, J., Columbia University
O'Neil, S.
Schnabel, N., NYU
Wang, Y.
Dai, M.
Singh, N.
Tu, R., City College of New York

Engineered protein hydrogels are an excellent resource for biomaterials development since they possess controllable pore sizes, flexible morphologies, and tunable mechanical properties. Manipulation of the protein secondary structure can accommodate diverse biological properties and associated applications. Protein-based block polymers have been used for the development of hydrogels with potential biomedical applications in drug delivery and tissue engineering. We have engineered a protein triblock polymer, CEC, comprised of two self-assembling domains (SADs): cartilage oligomeric matrix protein coiled-coil (C) and elastin (E) domains. CEC is capable of forming a thermoresponsive hydrogel with the ability to encapsulate small molecules into the hydrophobic core of the C blocks. At low concentrations, CEC assembles into a micellar-like protein particle induced by the E domain. As assessed though rheology, it demonstrates elastic character that is dependent on the C blocks and micelle packing. It exhibits reversible folding and unfolding as well as enhanced small molecule recognition and binding.

Synthetic Biological Platform of Corynebrick Vectors and Its Application to Improve Xylose Utilization and Cell Growth in Corynebacterium Glutamicum

ICBE Asia

Conference: International Conference Biomolecular Engineering ICBE

Year: 2018

Proceeding: ICBE Asia 2018: International Conference on Biomolecular Engineering

Group: Poster Session

Session: Poster Session

Time: Tuesday, January 9, 2018 - 4:05pm-5:30pm

Authors:

Park, J., Sungkyunkwan University
Yoon, J., Sungkyunkwan University (SKKU)
Woo, H. M., Sungkyunkwan University

Metabolic engineered *Corynebacterium glutamicum*, the non-pathogenic microorganism, is widely contributes to the industrial production of amino acid. To broaden its utilization of metabolic engineering in *C. glutamicum*, we established a synthetic biology platform vector, called CoryneBrick vector, which can be used as an expression vector in *C. glutamicum* and as an *E. coli-C. glutamicum* shuttle vector. For application of the CoryneBrick vector, we constructed the xylose utilization pathway encoding the *xylA* and *xylB* in the xylose-negative *C. glutamicum* wild type by rapid BglBrick cloning method, resulting in a growth rate of 0.11 ± 0.004 h^{-1} and a xylose uptake rate of 3.35 mmol/gDW/h when 1% xylose was used as sole carbon source. Also, we applied the CoryneBrick vector with a conventional *C.*

*glutamicum*expression vector for two-plasmid expression systems. Using this two-plasmid expression system, we improved xylose utilization of *C. glutamicum* not only introducing the xylose utilization pathway but overexpressing the pentose phosphate pathway genes *tal*, *gnd*, or the phosphoketolase gene *xpkA*, resulting increased the growth rates and xylose uptake rates compared to the wild-type with xylose utilization pathway alone. Thus, this study could be useful for metabolic engineering of *C. glutamicum* to produce amino acid and other high valued chemicals. This work was supported by the BK21 plus program through the National Research Foundation (NRF) funded by the Ministry of Education of Korea.

Reference

[1] Kang, M.K., Lee, J., Um, Y., Lee, T.S., Bott, M., Park, S.J., Woo, H.M., Synthetic biology platform of CoryneBrick vectors for gene expression in*Corynebacterium glutamicum* and its application to xylose utilization (2014) Appl.Microbiol. Biotechnol.

[2] Jo, S., Yoon, J., Lee, S.M., Um, Y., Han, S.O., *Woo, H.M., Modular pathway engineering of *Corynebacterium glutamicum* to improve xylose utilization and succinate production (2017) J. Biotechnol.

Transcriptome Analysis of Pyropia Yezonesis Compared with Heat-Stress Tolerance Mutant Using RNA-Seq

ICBE Asia

Conference: International Conference Biomolecular Engineering ICBE

Year: 2018

Proceeding: ICBE Asia 2018: International Conference on Biomolecular Engineering

Group: Poster Session

Session: Poster Session

Time: Tuesday, January 9, 2018 - 4:05pm-5:30pm

Authors:

Park, S., Chonnam National University
CHOI, J. I., Chonnam National University
Ngan, T. K., Chonnam National University

Pyropia yezonesis is one of the macro marine algae and is used to sea crop in the East Asia such as Korea, China and Japan. *P. yezonesis* is present mostly in intertidal zone. Due to the feature of the growth environment, it is received many kinds of stress during growing such as light, temperature, nutrient and dry. It is important to withstand these stress conditions for growing. In recently, many studies are tried to selecting pyropia strain had stress-tolerance. But, there are very few studies understanding molecular mechanisms of stress-tolerance mechanisms in pyropia.

So, we made the heat-stress tolerance *P. yezonesis* mutant stain using radiation and got whole transcriptomes using Next generation RNA Sequencing (RNA-seq) to study stress-tolerance mechanisms. RNA-seq reads were assembled using Trinity with *de novo* assembly mode and generated 31,147 and 78,793 unigenes for mutant and WT, respectively. For downstream analysis, the two transcriptomes were annotated by searching homologous sequences in NCBI NR, Swissprot, Pfam, and KEGG database. We analyzed differentially expressed unigenes and gene ontology (GO) enrichment analysis using shared sequences between WT and mutant. And putative Single Nucleotide Polymorphisms (SNPs) were discovered between two transcriptome sets (shared or unshared).

Plenary Speaker: Matt DeLisa, Cornell University: Microbial Glycoengineering Research

Conference: International Conference Biomolecular Engineering ICBE

Year: 2018

Proceeding: ICBE Asia 2018: International Conference on Biomolecular Engineering

Group: General Submissions

Session: Plenary Lecture: Matt DeLisa, Cornell University

Time: Wednesday, January 10, 2018 - 9:00am-10:00am

Authors:

DeLisa, M., Cornell University

Zhangyong Si[1], Yu Du[1], Moon[1], Damien[1], Jothy[1], Yang Liu[2], Kaixi Zhang[1], Sheethal[1], Chun Kiat Yeo[1], Yuguang Mu[2], Kevin Pethe[3], MB Chan-Park[1]*

1. Centre for Antimicrobial Bioengineering, School of Chemical and Biomedical Engineering, NTU, 62 Nanyang Drive, Singapore 637459, Singapore.
2. School of Biological Sciences, NTU
3. Lee Kong Chian School of Medicine, NTU

The rise of multi-drug resistant (MDR) bacteria together with the decrease of useful antibiotics has been a mounting problem. Especially for Gram-negative bacteria, no new family of drug was found in the last past 50 years. Herein, we developed a novel glyco-Î^{2}-peptides through one-step polymerization, following by one-pot global deprotection to get water soluble polymers. Our glyco-Î2-peptides had relatively good antimicrobial activity against both Gram-negative *P. aeruginosa* and Gram-postive *S.aureaus*, with MIC of 64 µg/mL. In vivo biocompatibility test using a murine intravenous toxicity model at dosage of 10 mg/kg indicated that glycopeptide had excellent biocompatibility, with 100% animal survival after 7 days and no significant difference in all the clinical important biomarker compared to the control groups. Our polymer could work synergistically with antibiotics at low concentration (16µg/mL), to potentiate conventional traditional antibiotic against wide type*P.aeruginosa* and their clinical relevant strain PAN-sensitive (PAES), clinical isolate P. aeruginosa D25, P. aeruginosa D1 and clinical multi-drug resistant strain PAER (almost no antibiotic showed effect except colistn) greater than 16 times.

Invited Talk: Engineered biosynthesis of functionalized natural products

Conference: International Conference Biomolecular Engineering ICBE

Year: 2018

Proceeding: ICBE Asia 2018: International Conference on Biomolecular Engineering

Group: General Submissions

Session: Session 5: Foundational Technologies for Biomolecular Engineering

Time: Wednesday, January 10, 2018 - 10:30am-11:05am

Authors:

Zhang, W., University of California, Berkeley

Natural products are important small molecules widely used as drugs, pesticides, herbicides, and biological probes. These compounds offer functional group arrays and scaffold architectures that form advanced platforms for optimization of biological activities. The unique functional groups, additionally, may be used as chemical reporters to enable the visualization, engineered biosynthesis and mode of action study of natural products. We recently identified and characterized novel biosynthetic pathways in microbes leading to the generation of bioorthogonal chemical entities that are commonly used in chemical synthesis and bioorthogonal chemistry. We further showed the promise of using these enzymatic machinery for *in situ* generation and incorporation of a unique chemical tag into various natural product scaffolds. Application of tagged natural products, such as *in situ* quantification and visualization, will also be discussed.

Formatting Issue - Best Available Paper/Graphic

Engineering a Fast-Responding Bacterial Test for Zinc Deficiency

Conference: International Conference Biomolecular Engineering ICBE

Year: 2018

Proceeding: ICBE Asia 2018: International Conference on Biomolecular Engineering

Group: General Submissions

Session: Session 5: Foundational Technologies for Biomolecular Engineering

Time: Wednesday, January 10, 2018 - 11:05am-11:30am

Authors:

McNerney, M., Georgia Institute of Technology
Styczynski, M. P., Georgia Institute of Technology

Zinc deficiency is responsible for over one hundred thousand childhood deaths each year, but the high cost and logistical challenges associated with current zinc diagnostic tools prevent adequate surveillance and treatment of zinc deficiency. To be used in nutrition surveillance programs, a zinc diagnostic must be inexpensive, easy to administer, nearly equipment-free, and fast-responding. A bacterial biosensor has the potential to meet all of these requirements. Our group recently developed an *E. coli* biosensor that produces different visible outputs based on the concentration of zinc in which the cells grow. However, these sensor cells always produce some pigment, which necessitates that they be grown from a very small inoculum so that initial coloration is minimal â€" thus requiring overnight assay times. To make an easy-to-use, field-deployable assay, we engineered cells that completely repress pigmentation during growth and quickly produce the appropriate pigment upon induction. This enables sample addition to concentrated pre-cultured cells, decreasing assay time from about 24 hours to less than 4 hours.

We first focused on eliminating unwanted pigment production. A sensor that produces pigmented metabolites (instead of more commonly used fluorescent proteins) is advantageous in resource-limited settings, since sensor readout can be interpreted without advanced equipment. However, using pigments as reporters is challenging, primarily because small amounts of unwanted enzyme can produce visible amounts of pigment. To combat unwanted pigment expression, we engineered three commonly used inducible promoter systemsâ€"*pLac, pBad,* and *pT7*â€"to be more repressible. Despite orders of magnitude decreases in uninduced protein expression, even the most repressible systems could not fully eliminate production of the red pigment lycopene. Translational modifications proved much more effective. When the original ribosomal binding sites for the lycopene-producing proteins were replaced with weak RBSs, all systems fully repressed lycopene, and upon induction produced visible lycopene within three hours. Supplementation of metabolic precursor pathways further reduced the time required for visible pigmentation to 1.5 hours.

We next incorporated zinc-responsive transcriptional elements into these systems to engineer cells that respond to different zinc concentrations by producing one of multiple pigments when induced. We added operator sites for the repressor Zur to inducible promoters to create hybrid promoters that respond both to an exogenous inducer (IPTG or arabinose) and to zinc. The location and number of operator sites was optimized, and the best-responding hybrid promoters only produce visible pigment in the presence of inducer and in low zinc concentrations. The zinc-responsive activator ZntR and its cognate promoter were then added to the circuit to control production of the protein that converts lycopene to beta-carotene. The resulting cells are colorless during the uninduced growth stage, and upon addition of inducer, they produce violacein (purple), lycopene (red), or beta-carotene (orange) to indicate low, medium, or high zinc concentrations, respectively.

The resulting biosensor is a significant step towards a field-friendly zinc diagnostic tool. More generally, this work demonstrates ways that synthetic biology and metabolic engineering approaches can be used to engineer systems that precisely respond to changes in their environment.

Formatting Issue - Best Available Paper/Graphic

Direct Conversion of Carbon Dioxide to Value-Added Chemicals Using Engineered Cyanobacteria

Conference: International Conference Biomolecular Engineering ICBE

Year: 2018

Proceeding: ICBE Asia 2018: International Conference on Biomolecular Engineering

Group: General Submissions

Session: Session 5: Foundational Technologies for Biomolecular Engineering

Time: Wednesday, January 10, 2018 - 11:30am-11:55am

Authors:

Woo, H. M., Sungkyunkwan University

Concerns about energy security and environmental issues has focused attention on metabolic engineering photosynthetic microorganisms to convert CO2 to value-added chemicals. Thus, I present the recent published research results of solar-to-chemical and solar-to-fuel using engineered cyanobacteria[1], including photosynthetic production of biodiesel (fatty acid ethyl ester), acetone, phytochemicals (amorphadiene and squalene)[2-4]. To achieve the goals using cyanobacteria, modular pathway engineering has been applied to reconstruct the metabolic pathways by introducing de novo pathway and re-directing carbon fluxes in cyanobacteria. Furthermore, a perspective of the photo-bioprocess engineering will be discussed to achieve the industrial goals of carbon utilization.

[1] Woo, H.M., Solar-to-chemical and solar-to-fuel production from CO2 by metabolically engineered microorganisms. Curr. Opin. Biotechnol. 2017, 45:1â€“7.
[2] Lee, H.J.; Choi, J.; Lee, S.M.; Um, Y.; Sim, S.J.; Kim, Y.; Woo, H.M., Photosynthetic CO2

conversion to fatty acid ethyl esters (FAEEs) using engineered cyanobacteria, J. Agric. Food. Chem. 2017, 65(6) 1087-1092.
[3] Choi, S.Y.; Lee, H.J.; Choi, J.; Kim, J.; Sim, S.J.; Um, Y.; Kim, Y.; Lee, T.S.; Keasling, J.; Woo, H.M., Photosynthetic conversion of CO2 to farnesyl diphosphate-derived phytochemicals (amorpha-4,11-diene and squalene) by engineered cyanobacteria, Biotechnol. Biofuels. 2016, 9:202.
[4] Chwa, J.W.; Kim, W.J.; Sim, S.J.; Um, Y.; Woo, H.M., Engineering of a modular and synthetic phosphoketolase pathway for photosynthetic production of acetone from CO2 in Synechococcus elongatus PCC 7942 under light and aerobic condition, Plant Biotechnol. J. 2016,14: 1768-1776.

Engineered Commensal Microbes to Target Colorectal Cancer

Conference: International Conference Biomolecular Engineering ICBE

Year: 2018

Proceeding: ICBE Asia 2018: International Conference on Biomolecular Engineering

Group: General Submissions

Session: Session 5: Foundational Technologies for Biomolecular Engineering

Time: Wednesday, January 10, 2018 - 11:55am-12:20pm

Authors:

Ho, L. C. L., National University of Singapore

Commensal microbes are known to regulate a healthy intestinal microbiota, translating to improved metabolism and immunity. Genetically engineered commensals have been studied for the treatment of inflammatory bowel disorders such as Crohnâ€™s disease, ulcerative colitis, as well as for diseases linked to hyperactivity of the host immune system. Using the commensal microbes as a chassis, it is possible to add health benefiting properties that can enhance the gut conditions of the host. Herein, we present our recent work using engineered commensal microbes to target colorectal tumor leading to tumor regression. A combination of two strategies is employed in order to allow effective targeted drug delivery to colorectal tumors by the engineered commensal microbes. First, the commensal microbes are designed to express cancer biomarker-binding proteins on the surface of the probiotic membrane for specific binding to colorectal tumors. Second, these microbes are engineered to secrete an enzyme that converts compounds derived from daily food consumption into anti-cancer compounds. The class of enzyme was screened to have stronger binding affinity to the substrate as well as improved pH and temperature tolerance for better performance in the gastro-intestinal tract. As the microbes attaches on the tumor surface, the enzyme is then delivered to the cancer cells a secretion tag. These engineered probiotics caused cell-line specific cancer cell death of up to 95% *in vitro* and led to latent tumor formation and reduced tumor size upon treatment in *in vivo* murine models. Our work demonstrates that our engineered microbes can potentially be deployed for enhanced for cancer treatment, leading to reduced colorectal cancer incidences and improved patient recovery.

Invited Talk: VM202, a DNA-based Potential Disease-Modifying Treatment for Painful Diabetic Neuropathy and Foot Ulcer

Conference: International Conference Biomolecular Engineering ICBE

Year: 2018

Proceeding: ICBE Asia 2018: International Conference on Biomolecular Engineering

Group: General Submissions

Session: Session 6: Translational Biomolecular Engineering

Time: Wednesday, January 10, 2018 - 2:00pm-2:35pm

Authors:

Yu, S. S.

VM202 is a DNA-based medicine designed to simultaneously express two isoforms of protein called hepatocyte growth factor (HGF). HGF is well known to induce the formation of new blood vessels and the growth and regeneration of nerve cells. Currently, VM202 is being tested for 4 major cardiovascular or neurological diseases. In all cases, VM202 is delivered by a simple intramuscular injection around the affected area. A phase III study for chronic, non-healing foot ulcer associated with diabetes (VM202-PAD) has been approved by the US FDA and initiated patient enrollment in 2017 in the US. A phase III study for painful diabetic peripheral neuropathy (VM202-DPN) was successfully initiated in the US in 2016, and more than 300 patients have been treated. Following the success of phase I trial for coronary artery disease (VM202-CAD), a phase II study is underway in Korea. VM202 also successfully completed a phase I/II trial for amyotrophic lateral sclerosis (VM202-ALS). In this presentation, the clinical development results of VM202 will be shared.

Biomolecular mechanism discovery from omics data: applications from clinical data to natural product chemistry

Conference: International Conference Biomolecular Engineering ICBE

Year: 2018

Proceeding: ICBE Asia 2018: International Conference on Biomolecular Engineering

Group: General Submissions

Session: Session 6: Translational Biomolecular Engineering

Time: Wednesday, January 10, 2018 - 2:35pm-3:00pm

Authors:

Eisenhaber, F., Bioinformatics Institute (BII) A-STAR Singapore

Penetration of omics technologies, especially genome and mRNA sequencing, combined with new imaging modes into the clinical environment creates unprecedented opportunities for understanding patient-specific aspects of pathogenesis and for designing individualized diagnosis and therapy approaches. Whereas there is great optimism that cost, logistics, infrastructure and data processing issues will be overcome in a close foreseeable future, the missing understanding of biomolecular mechanisms that translate individual genome aberrations and mutations into phenotypes will likely hinder educated, rational individualized intervention for many types of pathologies. To note, most of the human genome (~99%) including about half of the protein-coding regions remain functionally insufficiently or not at all characterized.At the same type, phenomenal progress will be seen when mechanisms are understood. Biomedical data analysis research at the Bioinformatics institute is aimed at contributing to closing this knowledge gap. Prioritizing of patient- and population-specific SNPs provides clues on the individualized pathogenesis and on the emergence of drug resistance. Omics studies of patient tumors are the key to elucidate driver mechanisms and to select efficient therapies if functions of mutated and mis-regulated genes are known. As smaller genomes of microbial and viral infectious pathogens are much better understood than their human counterpart, omics-based diagnoses can efficiently guide public health measures and individualized therapy. As microbes and fungi play a great role in biotechnology, the understanding of their genome-phenotype relationship has great impact on new secondary metabolite and enzyme discovery. Reference: "A decade after the first full human genome sequencing: when will we understand our own genome?" JBCB 2012;10:1271001

Engineering of Peptide Linkers for Heterogeneous Enzymatic Production of Chiral Amines

ICBE Asia

Conference: International Conference Biomolecular Engineering ICBE

Year: 2018

Proceeding: ICBE Asia 2018: International Conference on Biomolecular Engineering

Group: General Submissions

Session: Session 6: Translational Biomolecular Engineering

Time: Wednesday, January 10, 2018 - 3:00pm-3:25pm

Authors:

Caparco, A. A., Georgia Institute of Technology
Bommarius, A. S., Georgia Institute of Technology
Champion, J. A., Georgia Institute of Technology

Immobilizing enzymes is known to improve their stability and process lifetime.[1] Working with nano- and microscale immobilization matrices, it is possible to overcome mass transfer resistance that is often associated with macroscopic immobilization materials. The leucine zipper pair, $Z_E$and Z_R, allows for high fidelity and high affinity heterodimerization, and may be used to create an affinity immobilization system.[2,3] In our work, enzymes are recombinantly fused to Z_E with an amino acid linker to provide adequate spacing, while a protein-based calcium-phosphate particle is constructed with Z_R. The hierarchically structured porous calcium phosphate supraparticles serve as a modular support for enzyme immobilization.[4]
A glutamate leucine zipper (Z_E) was added to both a chimeric amine dehydrogenase (AmDH)[5] and formate dehydrogenase from *C. boidinii*(cbFDH) using a library of amino acid linkers varying in length, proline content, and terminal orientation. The linker candidate which

retained the highest activity for both binding and catalytic activity was identified for each enzyme. Only one linker construct for each enzyme exhibited activity similar to the unfused enzyme, and the preferred linker for each enzyme varied significantly. This highlights the necessity to create a variety of linkers when designing a new fusion protein to find the composition with the highest activity. The enzymes were immobilized in supraparticles through Z_E/Z_R dimerization, and retained greater than 80% of the soluble activity.

1. Bommarius AS, Paye MF. Stabilizing biocatalysts. *Chem Soc Rev Chem Soc Rev.* 2013;42(42):6534-6565. doi:10.1039/c3cs60137d.
2. Moll JR, Ruvinov SB, Pastan I, Vinson C. Designed heterodimerizing leucine zippers with a ranger of pIs and stabilities up to 10-15 M.*Protein Sci.* 2001;10(3):649-655. doi:10.1110/ps.39401.
3. Park WM, Yee CM, Champion JA. Self-assembled hybrid supraparticles that proteolytically degrade tumor necrosis factor-a.*J Mater Chem B.* 2016;4(4):1633-1639. doi:10.1039/c5tb01647a.
4. Park WM, Champion JA. Colloidal Assembly of Hierarchically Structured Porous Supraparticles from Flower-Shaped Proteinâ^'Inorganic Hybrid Nanoparticles. doi:10.1021/acsnano.6b01003.
5. Bommarius BR, SchÃ¼rmann M, Bommarius AS. A novel chimeric amine dehydrogenase shows altered substrate specificity compared to its parent enzymes. *Chem Commun.* 2014;50(95):14953-14955. doi:10.1039/C4CC06527A.

Application of Halophilic Archaea for the Production of Polyhydroxyalkanoates (bioplastics)

Conference: International Conference Biomolecular Engineering ICBE

Year: 2018

Proceeding: ICBE Asia 2018: International Conference on Biomolecular Engineering

Group: General Submissions

Session: Session 6: Translational Biomolecular Engineering

Time: Wednesday, January 10, 2018 - 3:25pm-3:50pm

Authors:

Mahansaria, R., Jadavpur University
Mukherjee, J.

Biodegradable plastics or bioplastics have raised great interest globally since the continuous decline of fossil fuel reserves used in manufacturing petroleum-derived plastics and their recalcitrant nature is slowly degrading the environment. Amongst the bio-based polymers, polyhydroxyalkanoates (PHAs) are accumulated as intracellular reserve materials by many microorganisms when carbon sources are present in excess and other nutrients like nitrogen, phosphorus, oxygen are limited in supply. PHAs are completely bio-synthesized and bio-polymerized, possessing over 150 monomer variations having different material properties. However, widespread commercialization of PHAs is still striving due to their higher production costs compared to that of synthetic plastics. While the price of polypropylene and polyethylene is US$ 1.32-1.91/kg, PHA cost is estimated as US$ 4.95-6.05/kg. In recent times, production of PHAs by halophilic archaea have gathered huge attention as they have many advantages like rapid growth using cheap carbon sources and waste products, less chance of contamination, no requirement of medium sterilization, re-use of medium salts, use of seawater as the salt-containing medium, thus reducing fresh water consumption and PHA recovery by easy lysis of cells with water. Here an overview of the work done in our laboratory is presented. Previous studies using ethanol industry waste, vinasse (molasses-based) and stillage (rice-based) as substrates to produce PHA from halophiles have been reported by our laboratory (Pramanik et al. 2012; Bhattacharyya et al. 2012, 2014, 2015). *Haloarcula marismortui* cultivated using 100% pre-treated vinasse resulted in 30% poly(3-hydroxybutyrate) (PHB) accumulation with 0.02 g/l-h volumetric productivity. Later, the study was improved using a more potent archaeon, *Haloferax mediterranei*, where 25% and 50% pre-treated vinasse were used to yield a more elastic biopolymer, poly(3-hydroxybutyrate-*co*-3-hydroxyvalerate) (PHBV) amounting 70% and 66% with productivity of 0.21 and 0.18 g/l-h, respectively. 100% thin stillage was later used by the same archaeon resulting in 71% PHBV accumulation with productivity 0.17 g/l-h. 96% medium salts of the spent stillage were recovered using a temperature-dependant directional solvent, decanoic acid. The production was upgraded in a 14 l simple plug-reactor producing 63% PHBV with productivity of 0.136 g/l-h and almost 99.3% medium salts were recovered. The cost of PHBV was estimated as US$ 2.05/kg when the annual production was simulated as 1890 tons. In addition, from our laboratory, Mahansaria et al. (2015) discovered 14 new haloarchaeal and halobacterial species as novel PHA producers from the salt producing multi-pond solar salterns in the states of West Bengal and Gujarat in India via a polymerase chain reaction-based screening method developed to detect the presence of the enzyme PHA synthase (*phaC*). The method was also tested on 9 type strains yielding 5 *phaC*-positive species. The method was developed as the existing techniques for detection of PHAs in environmental halophilic archaea/bacteria are either imprecise or require prior PHA production before screening. The newly developed method involved amplification of approximately 280-300 bp conserved region of Class III *phaC* gene of halophiles using primers codehopCF and codehopCR, thus eliminating the false positives seen with traditional Nile Red staining procedure and can be used universally to detect PHA producers.

AIChE
120 Wall Street, FL 23
New York, NY 10005-4020

ISBN 978-1-5108-9198-2

CURRAN ASSOCIATES INC.
proceedings.com